God's Plan for the Seashore

by
Judy Hull Moore

Edited by Delores Shimmin

Library of Congress Cataloging in Publication Data

Moore, Judy Hull.
  God's plan for the seashore.

  SUMMARY: Describes plants, animals, and terrain of the
seashore that God made.
  1. Seashore—Juvenile literature. 2. Nature—Religious interpreta-
tions—Juvenile literature.
[1. Seashore]   I. Shimmin, Delores.   II. Title.
QH95.7.M59   574.909 '46   80-17802
ISBN 0-8024-3066-X

MOODY PRESS ● CHICAGO

©1976, by
A Beka Book Publications
Moody Press Edition, 1980

BN: 0-8024-3066X

God made the seashore.

Play on the seashore
And gather up shells.
Kneel in the damp sands
Digging wells.

Run on the rocks.
Where the seaweed slips.
Watch the waves
And the beautiful ships.

—Mary Britton Miller—

"The sea is his, and he made it."
Psalm 95:5

God made the seashells.

I like to hunt seashells.
Here are some pretty shells.
They are empty now.
Once a little sea animal
        lived inside each one.

God made the sand dollars.

A sand dollar is a round,
    thin animal.
It is shaped like a silver dollar.
It eats seaweed.

God made the crabs.

A crab is a funny little animal.

He has ten legs.

Two of them
    are pincers.

His eyes are
    on stems.

When he runs,
    he runs sideways.

A crab likes to fight.

He likes to pinch.

If his leg is bitten off, he will
    grow a new one.

Courtesy Carolina Biological Supply Company

God made the starfish.

The starfish is not a fish.
But it lives in the water
    with fish.
Most starfish have
five arms.

Color the starfish.

God made the sea gulls.

The sea gull is a gray and white bird.
It has long, strong wings.
The sea gull likes to eat
fish and insects.
It is a good
fisherman.

Draw a fish in the water for the sea gull to catch.

God made the sea horse.

The sea horse is a small fish.
He wraps his tail
    around sea plants
    to keep from
    floating
    away.

Color the sea horse.

God made the sea.

The seashore is a place where
the sea and land meet.
The seashore can be rocky.
The waves beat on the rocks.
The waves make
a loud noise.

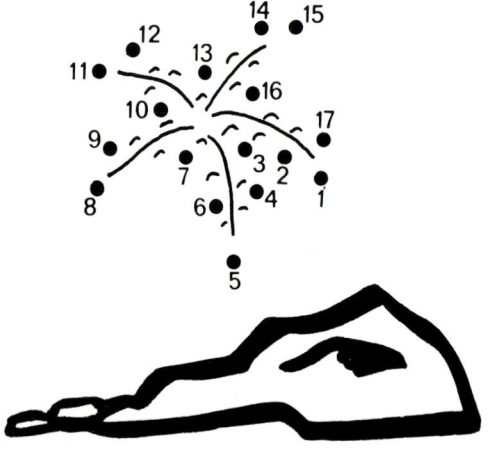

Follow the dots.

Follow the dots.

God made the sand.

The seashore can be sandy.
It is called a beach.
I can play in the sand.

Color me.

Follow the dots.

Thank you God, for the seashore.